Adeline Dutton Train Whitney

Zerub Throop's experiment

Adeline Dutton Train Whitney

Zerub Throop's experiment

ISBN/EAN: 9783337223625

Printed in Europe, USA, Canada, Australia, Japan

Cover: Foto ©berggeist007 / pixelio.de

More available books at **www.hansebooks.com**

ZERUB THROOP'S

EXPERIMENT.

BY

MRS. A. D. T. WHITNEY,

AUTHOR OF "HITHERTO," "PATIENCE STRONG'S OUTINGS," "THE GAYWORTHYS," "FAITH
GARTNEY'S GIRLHOOD," "WE GIRLS," ETC., ETC.

—o○;●;○o— -

L O R I N G, Publisher,

COR. BROMFIELD AND WASHINGTON STREETS,

BOSTON.

ZERUB THROOP'S EXPERIMENT.

I.

HOW ZERUB LEFT IT ALL TO PROVIDENCE.

ZERUB THROOP sat in his spring-lock sanctum.
It was a contrivance of his, whereby it might never
be precisely known whether he was out or in; also,
no other person, curious or dishonest, could invade
the place to occupy it even for a moment, except
with door carefully set wide. He carried the key
in his pocket. Once swung to, the heavy leaf fastened itself instantly; then he and his cigar and his
black cat were walled up together. Zerub always

3

kept a black cat. He had had six generations of them, all precisely alike. Where the type varied, the kitten was drowned.

A staircase led down from the passage without to the side entrance of his house. People on errands, or with bills, or to pay money, or receive orders, came here. Zerub could see from his window whom it might be.

He had an office directly below, where he made payments, and signed receipts, and gave such other audiences as he chose, holding thus pretty much all his limited intercourse with his kind. Unless he owed a man, or a man owed him, or one or the other wanted for money, money's worth of use, property, or service, what should there be between them? Zerub Throop always wanted to know that.

He had a little dining-room beyond his office. His sleeping-room was within his sanctum. What

if he should die there some night with his oak
sported?

The whole front of his large old house, a place
he had taken a whim to buy furnished as it stood,
was unused.

He had his head out at his window at this
moment at which we take him up. He was watch-
ing a woman who had come to the door below with
something to sell. She had come from a good way
off, peddling her wares, or she would never have
climbed Throop Hill.

"Tell the mistress it will be sure to make the
hair grow, if it's gone ever so."

"It isn't a mistress, it's a master," said the
servant Sarah, from within. "And he don't buy
hair-grease; and he won't have peddlers.".

"It isn't grease; it's Phœnix Regenerator.
It'll—"

"It's no use, I tell you. Not if it would save

souls. I tell you he don't buy things." And Sarah, bethinking of her half-ironed shirt-bosom, and her cooling flats, shut the door summarily.

Zerub Throop laughed. The woman looked up.

"My hair never comes out, madam, I assure you," said he with a mocking blandness, and a half bow of his thickly-covered, close-trimmed, grizzled head. "I'm not in the habit of losing things."

"You might, though," she answered, as ready as he. "You might begin; and it's things that never went before that goes worst if they once sets out. When it once begins to drop, you'll — "

"Hammer it in, ma'am! and rivet it on the other side. Good-morning;" and Zerub shut his window.

"Hammer it in! I guess you're used to hammerin' in; feelin's and Christian charities and such. Done the undertakin' business pretty much all

along, I should say. Well, wait till *you're* hammered in, and riveted on the other side!"

As she walked out of the upper gate upon the
hill, another woman rang the bell at the front-door.
The sound pealed through the house startlingly.

Hardly once in a year did any one ring at Zerub
Throop's front-door. One had to turn aside from
the gravelled drive to reach it, across a grass-plot.
Old vines, little trained or cared for, tangled up the
porch-way; but Mrs. Whapshare came to the front-
door. She had been ten years making up her mind
to come at all, — ever since her husband died, and
left her poor. Now her little children were growing up, she had a hundred needs for them to-day
that pressed her sorer than the needs of ten years
ago. They might go out into the world to make
their way; but she wanted life-tools to give them to
go out with. Training, knowledge, opportunity, —
these things, in the outset, must always cost some-

body something. She could not give them bread and butter now, and send them to bed. There was other feeding that they were hungry for.

Zerub Throop knew Mrs. Whapshare by sight, as he knew nearly every man and woman in the town; but he had never spoken to her. Why should he? She was no tenant of his. He wanted nothing of her; she could buy nothing of him. The human relation, as Zerub understood it, failed. The wires were down.

Yet Mrs. Whapshare came, and rung at his front-door.

"There is a lady, sir, in the north-east room, askin' to speak to you," called Sarah, from outside the oak, not knocking, for she knew now that he was there.

"Why didn't you get rid of her, as you did of the Regenerator?"—half pleased, half surly, at her management; first good, then bad.

" She isn't the regeneratin' sort. She aint got bottles, nor yet books, nor yet fortygraphs of President Grant and Mr. Bismarck Brown. There aint nothin' to send her off *on*. She jest wants to see you. I can tell you who 'tis. It's Mis' Whapshare, down Ford Street way. She stepped in as if she'd made up her mind ; and it's one of the little ones that makes up with a twist."

Sarah Hand was almost the only person who ever made many words with Zerub Throop ; but her words suited and amused him, and she knew it. It was with a sort of crusty good-humor that he went down into the dim and musty north-east parlor, where Sarah had folded back a single shutter, to see Mrs. Whapshare.

The lady rose as he entered, stirring the gloom and must of the corner in which she had seated herself, and gathering up, as it were, the darkness into shape with the shadowy movement of her black dress.

Zerub bowed.

"Mrs. Whapshare," said the lady, — "Mrs. Miles Whapshare."

Zerub sat down, and waited for more.

"I have come to ask you something, Mr. Throop."

"Of course, madam. They all do," answered Mr. Throop politely, drawing down his waistcoat, and leaning back in his chair, laying his right foot across his left knee, and folding his arms, as a human being in a state of siege instinctively barricading himself.

Mrs. Whapshare looked at him quickly. She changed her tone and approach. She was not a timid woman, though she had been ten years making up her mind.

"I beg your pardon, sir, I began wrong. I mean, I came to *tell* you something."

Mr. Throop bowed.

"You owed my husband, Miles Whapshare, fifteen thousand dollars."

"Once I did," answered Mr. Throop.

"Don't you think — I mean *I do* think — you owe his children something now."

"In this country, madam, no one is persecuted for opinion's sake. You have a perfect right to think so, — and — to continue thinking so."

Mrs. Whapshare was forced back to her questions. "Don't *you* think so, Mr. Throop?"

"No, madam. I am quite willing to answer any inquiry you would like to make. I do not think so."

Mrs. Whapshare had to put it interrogatively again. Otherwise, it was plain the conversation

was to drop, and in like manner would perpetually
drop.

" Why, sir ? "

"In the first place, madam, three and twenty
·years ago Miles Whapshare hadn't any children.
Whatever responsibilities he undertook afterward,
he undertook in the face of his business loss. He
began the world again, as I did. *I* couldn't afford
children, ma'am. In the second place, I paid him,
as I did everybody else, twenty-five cents on the
dollar, and was discharged. I began again, and
worked up. If Miles Whapshare didn't work up,
that is simply the difference between us. In the
third place, if I were to call it a debt now, how
much do you think the debt would be ? "

"I don't know. I don't know as that alters
it."

"I'll tell you, then. Upon fifteen thousand
dollars, I paid Miles Whapshare three thousand

seven hundred and fifty, leaving eleven thousand two hundred and fifty. That, at simple interest, would by this time just about have increased by one and a half. Do you think I owe Miles Whapshare's children to-day twenty-eight thousand one hundred and twenty-five dollars? It is either that or nothing."

"I think it is likely it is that, then," replied Mrs. Whapshare, with a calm indifference to the figures. "But they would be glad of a very small proportion."

"Possibly. Miles Whapshare was. But you leave the argument. The grandchildren might come back with their claim, by and by. The world doesn't go trailing on after that fashion. When things are squared up, they are squared. There had to be a deluge, once, ma'am, and the race began again. Pope Gregory had to strike ten days out of the year 1582, to bring the world's account down to what the sun could pay; and I believe you think your sins are settled for on much the same principle, don't you?" Bankruptcy and discharge seem to be taken into the original plan of things. At any rate, that is what occurs, and there is an accepted order for it. Is this all, madam? and is your mind satisfied?"

And Zerub Throop arose.

The woman's figure in black moved again also, making that shape of shadow in the gloomy sofa-corner. A voice that trembled now came out of the shade.

"It seemed to me as if it ought to have been, somehow; a few thousand dollars would have been so much to us all this time! and I knew you owed it once. You are rich, Mr. Throop; and you have nobody to keep your money for."

"I can leave it to cats and dogs if I like. I can do as I please with my own."

"You may think you can," said the widow, speaking firmly again; "but it will be as Providence pleases, after all. Even the king's heart is in the hand of the Lord."

"Very well! Try Providence; but, if Providence is anything like Zerub Throop, it won't do to begin by telling him he owes you an old debt on somebody's else account."

"You know about that Mrs. Whapshare?" Mr. Throop said, interrogatively, to Sarah Hand, when she was bringing in his dinner, — a roasted duck, with port-wine sauce. "She's a pretty comfortable sort of person, I should think."

"Well," answered Sarah, "folks *is* most alwers pretty comfortable, aint they, 'xcept the regular give-up starvation ones? You see 'em goin' 'round; and they has shoes an' stockin's on, an' gowns, an' bunnits, or coats and hats; an' they goes in somewheres when it rains, or it comes night; an' they git breakfast, an' dinner, an' supper, I s'pose, or else they wouldn't *be* goin' 'round. You don't see 'em droppin' nowheres. Of course, they're comfortable. Everybody gets shook down into some sort of a place. 'The world's like a hoss-car; they git in, an' they git out; an' they've been took along between. Some sets down, and some stands up, and some hangs on to the straps. Some gits into a place at the beginning, and some slips into one when somebody else gits out. There don't seem to be no rule about it; it regilates itself."

"But Mrs. Whapshare? — she lives in a good house."

" They can't eat shingles and timbers, though. 'Taint like little King Boggins."

" She has a roof over her head, however, and it is her own. She has several children."

" More. She's got six."

"All grown up?"

"Well, the everidge of 'em is. Charlotte, she's eleven. Miles Whapshare died ten years ago, and didn't leave much of anything but the old house and the garding and the six children and a mess of old store-books full of bad debts and tribulations."

"Been to school?"

"Children? Yes an' meetin', an' Sunday school, right straight along. John, he's got a place in a store. They're nice folks enough. Mis' Whapshare aint got much force to her, though."

"I should think she had done pretty well under the circumstances."

"That's just it. She's a woman that's always been *under* a lot of 'em, — clear down. What business do folks have to be under the circumstances, I wonder? Why don't they get on top of 'em? What is circumstances made for?"

"To *stand round*, Sarah," said Mr. Throop, in

2

italics. "If you knew Latin, you'd see. That's
what we've got to do with 'em. Keep 'em in their
places. Make 'em stand round!"

"Or *git*," said Sarah, sententiously.

Mr. Throop laughed.

"Bring me a lemon," he said; and Sarah, having
done that, understood that the conversation was at
an end, and withdrew, like a circumstance, into the
kitchen.

The one course over, Zerub went, as was his cus-
tom, upstairs to his wine, his dessert, and his
cigar. He never ate pastry. A little fruit was
set upon the round table, in his sanctum, also a
basket of small sweet biscuits, — these more especially for the benefit of the cat, to whom he fed
them; beside these, a bottle with cap of tinfoil over
the cork, his cigar-holder, tray, and match-box.
In this company, Mr. Throop always read his papers
after dinner for an hour. The cat, when she had

got biscuits enough, dozed beside him on a soft square sofa-cushion, flung down, for her use, upon the floor. Zerub pulled her ears once in a while, and woke her up to tell her the news, and what he thought about it.

"She knows, and she don't contradict," said he.

To-day, he did not read long.

"They'll get into a nice mess in Europe; won't they, Tophet? They've got to, sooner or later; that's what I told the Whapshare woman. The world's never safe from a muddle but when it's just out of one; and, if you can't be safe then for a while, what's the use of the muddle? Hey, old cat?"

Tophet rose lazily, stretched out her fore legs to their farthest possible extent, stretched up her hind ones, lifting her back into a heap, and dropping her neck into a hollow; then gathered herself together again, with raised and vibrant tail, and rubbed and coiled herself round her master's ankles.

"I wonder how it would seem to do it, old cat? I wonder what she would think herself, if I really did? See here, now;" and Mr. Throop drew forth his great wallet, and therefrom took a slip of white paper, such as he kept ready for bills and receipts. He dipped a pen into an inkstand that stood upon the table, and wrote four lines.

"That would do it."

He was only thinking now, not soliloquizing. Mr. Throop never did that foolish thing; he only talked out now and then, in scraps, to the cat.

He sat holding that which he had from a queer impulse written, fancying queer what-ifs about it.

"That would do it. Give that woman this slip of paper, and it turns her life right over for her, t'other side up again, — the side she hasn't seen for ten, twenty years, perhaps, by that time, no, nor ever; and it alters six lives after hers.

"I don't suppose anybody ever wrote exactly such a note as that; couldn't be discounted. It would stand good, though, when the time came. Mrs. Whapshare, two things are between you and this slip of paper, — my will, and my life. I can, and I can *not*. There comes in free agency, and all the rest of it. It is certain that I either shall or shall *not* turn this freak into fact. Certain somewhere. Where? In time, or Providence? Providence *may* meddle with such things; but I never came across Providence amongst 'em, that's all. I've had my way to work up; and I've been left pretty much to myself; and I've worked it. I'm left to myself now. Am I though? How do I know?

"See here, what if I do neither? What if I leave it to Providence to finish it, if it will?"

There was a small blank in one of the four lines.

Zerub Throop dipped his pen again, and filled the space with two words. He turned it over, and indorsed it with a date and a sentence. Then he laid down the pen, and sat folding and rolling the paper abstractedly several minutes until he held it in a tight round, like a very small Catherine-wheel, between his finger and his thumb.

"Would it ever fire off?" he wondered.

In the same whimsical, half-voluntary way, as if letting his vagary, that he might stop at any point, run on with him, he tore a bit of tinfoil from the sheath that had covered his bottle, and rolled it again, carefully and compactly, in that. He folded and pressed and smoothed the foil around it, and welded it into a silvery ball.

"Did you ever see a secret, Tophet?" he said to the cat. "That's a secret. That's the sort of thing it is, when you take it out of your mind, and look at it."

Then he sat holding it again, amusing himself so, — playing passively, as it were, with fate and possibility, — others' fate that he thought he held ; first in his own mind and will, — now that he had taken it out, and looked at it, between his thumb and finger.

But what was he to do next, or not to do, seeing he had given it up to Providence ? Providence would neither put it by, out of his thumb and finger, nor throw it away.

"I won't destroy the thing," he said. "I'll go as far as that, and then it is out of my hands. I'll leave it loose on creation. Things have to go somewhere. What difference will it make to me?"

He laid it out of his fingers, on the table, — anywhere, as it happened to fall.

"That's all between you and me, Tophet," he said.

"Ni — ai — o !" answered the cat.

"And — the post, Tophet; you and me and the post. What do people mean by the post?"

Then he took his hat and cane, and went off for his afternoon walk.

Zerub Throop was not an ill-souled man; he was only a strange, solitary one, — grown selfish and one-viewed through solitariness, and through having "worked his way up."

Sarah Hand came upstairs, found the door hooked back that she might enter, carried off the empty bottle, the fruit-basket, and the torn bit of tinfoil that was evidently rubbish, beside it. She picked up the round bright ball, looked at it, turned it over, saw that it was folded, not crumpled, and laid it into the little grooved lid at the top of Mr. Throop's writing-desk, to keep company with an old knife, a bit of sealing-wax, some used pens, and a piece of India-rubber. Sarah Hand never "cleared up" anything that could by any possibility ever be called for or thought of again. There were old bits of paper, scribbled with temporary calculations, tucked between the leaves of his blotting-book, thrust into his match-box, and clasped among the notes and scraps in his little gilt finger-clip, that had been dusted over and replaced for month after month, even year after year.

So, when Zerub came home, there the secret lay, taken care of by Providence and Sarah Hand. There it continued to lie for several weeks; till, one day, when he lifted the grooved lid to find something that was underneath, the silvery ball rolled out at the end, and upon the table, and down to the floor.

Zerub looked at it. "It's out of my keeping," said he; "I've nothing to do with it." And he let it lie.

Sarah Hand picked it up when she swept next day, and dropped it into the bronze match-box, where it fell to the bottom, among some stray tacks and screws and buttons that were safe there from being lost or wasted, and also from ever being drafted to any earthly use.

Zerub did not ask for it, or look for it. It had fairly got beyond his knowledge now, as when one wilfully loses count of some sound or motion one has pained one's self involuntarily in following, and is thankful to let go. One night, months after, he upset his match-box in the dark. The dust that fell from it got brushed up in the morning, the tacks and screws and buttons put back again, and nobody, of course, thought of or recollected anything more; until, that same afternoon, sitting with his wine and his paper and his cigar, Zerub saw the cat claw something from under the edge of the low, broad base of his round table, give it a pat, to try if it had life and fun in it, and send it shining across the floor.   .

"Why, that's — " said Zerub; but before he came to the exclamation-point at the end of his sentence, Tophet was after it again; and a second buffet drove it straight before his eyes to the one possible spot where it could get lost out of that room, — down the open lips of the old-fashioned, brass-valved register.

"That's all!" said Zerub, with a deliberate period. "Nothing is lost while you know where it is. But it's none of our business; is it, black cat?"

They two knew; and they never told.

Afterwards, Zerub Throop lived on for the space of two years and five months, and gathered to himself his interests and his dividends, and smoked his cigar daily after his dinner; but he never spoke again with Miles Whapshare's widow, or put her name again to any paper that he wrote or caused to be written; and at the end of this time, suddenly, and in the midst of his strength, he turned away from all these things, as if he had never striven for or possessed them, and went, as we all go, to "work his way" up farther.

II.

HOW IT WAS WITH THE WHAPSHARES.

MRS. WHAPSHARE went out through the tangled porch, and heard Mr. Throop draw the rusty bolt behind her. There was an odd blank in her mind as she walked down the hill into the town again, as if she had taken some hope up there with her that she had been long used to, and had buried it, and was coming back into her life alone, without it.

It had been, all these ten years, a kind of vague assurance to her to see Zerub Throop go by, up and down the street, and to think to herself, "That man failed, and owed my husband eleven thousand dollars that he could not pay. He has got it now, and plenty more; I've a great will to go, some day, and remind him of it."

It helped her, — this undefined hope and half-intent, — almost unconsciously, through many a hard pinch. She had a nut that she might yet crack, as they do in fairy tales, when they get to the worst; and who knew what might come of it? Anything, everything, might; and, so long as there is a "might" in one's life, one can go on; it is a reserve in the army of one's forces.

This morning she had gone and cracked her nut; and there had come out of it black ashes.

She looked so tired when she came in, that Martha, her daughter, did not tell her that the soup was burned; but she smelled it, coming in out of the fresh air. Burnt peas are pungent.

" There's our dinner gone!" said she.

" No," spoke out Caroline, from the kitchen; and she opened, with a gay clatter, the oven door.

" Smell my potato puff; and we've an omelet just ready ; and you're to have a cup of tea, with a table-spoonful of cream that I got off the bowl for you this morning."

That was Caroline Whapshare's way with things. Martha took them harder.

"I think the soup is always burned for us," she would say. "There's a wrong somewhere, that things should be so."

She was like the Jews, who asked, " Who hath sinned, this man or his parents ? "

Caroline had the Christ-answer ready.

" Not so much a wrong, may be, as something to be set gloriously right. How good it will be when the sun breaks out in the west, Mattie ! "

" Yes, away down; just a strip for the last minutes under the clouds, when the day is all gone."

"Even then, it is not as if there were not another coming."

"That does not help the Johnnie feeling."

Now, when John Whapshare had been a little boy, he had given the household this compound substantive and a proverb. They were trying to comfort him, for a childish disappointment, by telling him of the good time he was to have next week, at Thanksgiving. "Ye-e-s," he persisted, sobbing with undiminished vigor; "but what kind of a time be I a-havin' now?"

Martha thought the family had been brought up on the Johnnie feeling.

"Mother has lost something," she said to Caroline, over the dinner dishes, that day. "She looks as if she had had something put away, and had gone to get it, and it was not there."

"What queer ideas you have, Mattie!"

"Maybe. I feel all sharpened up, as if I knew

things through the ends of my fingers. Queer
ideas come of queer living. What are we going
to do with that old straw matting for winter?"

"It was rather a pity in the beginning. Children
do scrape their chairs so!"

"Well, it's the end now; and it has only lasted a
year. It is terribly expensive to be poor, Car. If
we had had a good ingrain for half as much again,
it would have lasted six years."

"I'll tell you what I have thought of," said Car.
"That north-east parlor, — we cannot do much with
it in cold weather. What is the use of having a
best room when you cannot have an every-day one?
We are right on the corner of the street; we might
let it for fifty or sixty dollars a year; and then there
would be the carpet and all the things to spare.
We could fill up with them splendidly for ever so
long."

"That very best Brussels carpet?"

"Well, yes; twenty-two years old, is it not? Older than either you or I, Mattie; which is all the reason we venerate it so. It was the best when we were born; and we were never allowed to have any crumbs over it. It is not handsome."

"But let a room? Who to, or what for?"

"To some comfortable old maid; or for an office, or a shop, or anything. Why should we care? I believe I shall put it into mother's head."

"How we should miss it in summer! — our only cool, shady place!"

"It is a good thing to let things go when you do not miss them. Then, when the missing time comes round, you rub along somehow. That's the way for poor folks to give. I've something else to propound, Mattie, some time; and I don't know whether to do it all in a heap, or to wait another year. For it must be a winter-strained notion too."

"I think when you are pretty well thumped

3

already is the time to take another. You might as
well keep on hammering."

" We might — sell — our — garden — for fifteen
hundred dollars, Martha Whapshare ! "

The first few words came slow and hard, trying their
way as they came, Caroline's eye fixed closely upon
Martha's face. The last all ran together in a great
hurry and triumph.

"We might — all get into our — caskets ! " answered Martha, with a sepulchral indignation.
"You would leave us just about room enough."

"Lydia ought to have those organ lessons that
she wants so much, and an organ to practise on. It
would be a profession for her."

"How do you know ? "

Caroline opened her eyes at her sister. "Why,
of course it would. Are they not building new
churches everywhere, all the time? and are not all
the women taking to preaching, which will leave a

capital chance for anybody that is willing just to glorify at the other end, without being seen of men?"

"Pshaw! I don't mean that. How do you know about the garden?"

"I asked Rufus Abell. He knows. I wouldn't go at mother, and stir her up for nothing, you see."

Martha rubbed the cover of a potato-dish silently for a full minute, looking at nothing, with that "setness" in her features, — her eyelids fixed at half-mast, neither lifting nor falling, a white pinch in the end of her nose, and the corners of her mouth crowded down with the close shutting of her small jaws, — as if her indignation at life were held in somewhere behind her face, as a smoker takes in and holds tobacco-smoke.

"She held her breath, and the mad went out at

her ears," she said once of herself when she was a child.

"I think it is a very prettily managed world," she remarked quietly, when she had put the dish-cover down, and shaken out the towel. " All Oregon and Alaska empty at one end, and people crowded out of their door-yards at the other. I'm going to talk to mother about it."

While " the mad went out at her ears," Martha's mind was always calmly made up to the inevitable. Her mother had lost some might, could, would, or should, to-day ; she had seen that ; she might as well piece out the conditionals for her. Martha Whap-share said her mother lived in the conditional mood.

Caroline knew how it would be beforehand ; it was the regular circumlocution of things in the family. She had the ideas. Martha growled at and presented them ; Mrs. Whapshare laid them up

among the mights, coulds, woulds, and shoulds;
now and then one was drawn out in an emergency,
and acted upon.

Rufus Abell came, and measured the garden-
piece. Rufus Abell was surveyor, real-estate agent,
broker, lawyer, executor, what-not, to half the
people, living or dead, who had, or had had, interests in Rintheroote.

There were thirty-two hundred square feet: "it
would sell," he said, "for fifty cents a foot; that
would be sixteen hundred dollars." Mrs. Whapshare went to bed with sixteen hundred dollars in
her pocket of possibilities. On the strength of that,
they had sirloin-steak for dinner the next day.
That did all the family good; in regular turn, it
would have been salt fish, — "One of the make-
believe days," Martha called it; when the dinner
was got over, and no one dined. They made believe, at regular intervals, with salt cod, baked

beans, pea-soup, and liver. That left three days
in the week for something real, — two at first-
hand, and one warmed up.

Mr. Abell also put a notice up at the post-office,
and into the village paper, of a desirable corner-
room to let in a dwelling-house, in a central locality,
suitable for a single lady or a professional man;
apply to him.

A great many people applied, — two washer-
women; a horse-car conductor with a wife and
seven children; an intelligence-office keeper; the
teacher of a boys' private school.. At last a young
doctor, newly come to the neighborhood, Arthur
Plaice, got it; paid twenty dollars in advance for
the first quarter, twelve of which Caroline Whapshare took to the city the next day, and paid, also
in advance, for the same length of time, for a Mason
and Hamlin organ. This came out on the same

express-wagon that brought Dr. Plaice's desk and
arm-chair and book-shelves.

They got acquainted with their tenant over the
unloading and bringing in. The ladies Whapshare
had been rather shy of him before.

He helped the express-man bring in the great
box into their sitting-room; then he stayed, and
unscrewed it for them, and drew the instrument
safely out, according to directions; then, when
they opened it, and wondered how it would sound,
and what Lydia would say when she came home,
he put a chair before it, and seated himself, opened
the stops, and touched the keys with a few beautiful
glad chords, and played what Caroline always
called afterward, the " Which being interpreted."
It had in it struggles and changes, and snatches of
comfort, and little climbing-up-hill notes, and sure
high ones, and droppings and sobbings down again ;
yes, and " the very little pinches too, that nobody

noticed but the pinched people;" and it had the great reach and longing; and, at last, a grasp and a joy, and a gentle flood of bright content, that filled the room and all their hearts as they listened, just as the sunset and the home pleasantness filled it, and glorified its new aspect; with the best things brought in for every day, and the "real Brussels," faded though it might be, on the floor, and the organ standing in the shady corner. .

The old maid, Miss Suprema Sharpe, lived right opposite, and could see, over her blinds, all that occurred. What she did not see, she heard; and, what she did not hear, she imagined; and what she saw, heard, or imagined, of a morning, for example, she ran up street, of an afternoon, and told to her friend, Mrs. Benny Dutell, while it was warm; just as she might carry ginger-cakes.

She was not a bad old maid, either; that is, she did not mean to be. She only lived all alone,

and there did not much happen to her. Nine
from four you can't; so you borrow ten. Miss
Suprema went borrowing ten all along the line.
She got things mixed up sometimes, and her sums
wouldn't prove.

Mrs. Benny Dutell was the postmaster's wife;
what came to her never grew cool in her hands;
so that you had your own story passed round to
you again presently, or even beforehand; as if it
had got ahead of the sun round the world, — by
the way of Upper Five Corners, or Lower Green
Point.

Dr. Plaice had hardly gone away into his office,
when Miss Suprema came "perpendiculating"
over. She walked very stiff and straight and
quick; so that she seemed like a stick shot broad-
side, instead of endwise, keeping its uprightness
as it went; or as a water-spout or a sand-column,
that slides tall and swift from horizon to horizon,

without a motion or a swaying, save determinately
on.

Nothing prevented Miss Suprema from getting
over sooner, and meeting Dr. Plaice there, but an
"*embarras de richesses.*" She stood in the middle
of her bedroom, and fairly spun when she saw the
furniture going in, and the big box, marked " Cabi-
net Organ," slid over the threshold along a board ;
when she spied, by the strong western light shining
in level through the room, the busy group about it
unpacking ; and when Dr. Plaice sat down and
began to play. Her bonnet was in the closet ; and
she would have to turn her back, and disturb her
hearing, to fetch it and put it on ; besides, if she
did, — which way? She was in a hurry to get to
Mrs. Benny's before the sun went down upon her
pheese ; and she was eager to gather more to go
with to-morrow. She wanted to run right in among
the Whapshares, and she did not want to · "stop

things;" the end was, that she came in upon their comfortable twilight complacency, waiting for Lydia's return and rapture.

"Well, I declare! You *are* spread out!"

Miss Suprema looked round the room beamingly. She looked at the carpet, and the gray moreen curtains, and the marble-topped pier-table; she did not mean to see everything all at once; she let the organ wait in its shady corner.

"No, Miss Suprema," said Caroline; "not spread out; only drawn in. The syrup is boiled down, that is all.

"To a richness? Well, how elegant you do look! You won't let it make any difference towards me, will you; but I may run in neighborly just the same, if I rub my feet well?"

Miss Suprema had quick little looks, that she sent everywhere out of her round brown eyes like a squirrel's; never moving her body, that sat straight

up from the edge of her chair, but only her head.
Lydia Whapshare said all she wanted was a bushy
tail, and a nut between her forepaws. But, to do
her full credit, the nut was seldom lacking meta-
phorically; and the tale was bushy enough by the
time she ran up the road again with it, along under
the wall.

With her swift, continued peeps, she was the
first to see Dr. Arthur Plaice, standing again in
the door-way of the room in the increasing twilight.

"Can you lend me a hammer for a moment,
Mrs. Whapshare?" he asked.

And while Mrs. Whapshare went for the hammer, Suprema Sharpe had a good look at him, with
what light there was at her own back, and full in
his face.

He was a very handsome man, she saw that, with
a square, firm figure, not over tall, a calm equipoise

in look and attitude, and all the indescribable bearing of a gentleman, that shows itself whether he stands quietly waiting, or moves and speaks.

He neither came into the room, nor withdrew shyly, but simply stood where the last natural act left him, until it should be time for the next. Self-consciousness, which is neither ladylike nor gentlemanly, always has to do something between. Dr. Plaice could make a pause. When Mrs. Whapshare brought him the hammer, he thanked her, and turned away.

"So that's him?" said Miss Suprema.

"That is Dr. Plaice," replied Mrs. Whapshare.

"Young, isn't he?"

"I dare say. I do not know his age."

"Just beginning. Well, you won't be much knocked up nights yet a while. To be sure, he's got the little east door to himself. It'll be sociable evenings. It's a good plan to have somebody there.

I wonder you never thought of it before. You
didn't really want that room. If you had only
made up your mind last year, there was little Lot
Green looking everywhere for a place to put up his
sign, and begin turnin' at law. You wouldn't have
had much company of him, though, for his evenings
were spoken for; and it wouldn't have been perma-
nent, because he's married now, and keeping house
and office all together. I guess it happened right
as it is."

"We had only just come through to the bare
floor here," said Martha, bluntly; "and I don't
suppose we shall have much to do with Dr. Plaice's
evenings."

"He's right in the house, anyway; and there's
always hammers and things; you'll get acquainted.
Well, I must go. I only looked in for a minute.
I'll come again. If anything should happen that
I shouldn't be able to come, you know, why, there's

the doctor; and one of my little quinsies might be an encouragement to him."

She fairly forgot the organ, after all.

She stood on the sidewalk for a moment, when she had got out, with a flapping in her mind that she was subject to, like a sail in a flaw of wind. She trimmed her decisions, however, quickly, and laid her course direct for Mrs. Dutell's.

She must go, sundown or not. She had a little joke on the tip of her tongue that tingled. Keep it over night? She might as well have tried to keep a Spanish fly there.

She was in too much of a hurry with it, though, when she reached Mrs. Benny's.

"It's easy enough to guess now what will *take Plaice!*" she cried right out, without preface.

"La! what?" said Mrs. Benny Dutell.

Then Miss Suprema saw that she had begun at the wrong end of her little joke, and spoiled it. I am viciously glad she did. I am **glad** she found out once in a while, in her own small way, which was all the way she could, how good it is to have things tipped out in a hurry, wrong end foremost.

There are two kinds of gossip, — the one that purely invents or recklessly misrepresents; and the one that shrewdly spies, puts this and that together, guesses, and anticipates; and the latter is indescribably the most aggravating. It was Miss Suprema's sort.

You can sit in your own room complacently, with a three weeks' influenza, and be told from outside that you have got the varioloid, or a softening of the brain; or that you have quarrelled with your wife or husband, and run away. All that will right itself; but to be informed that you are about to give out invitations to a party, or publish a book, or go to Europe, when you can't say you haven't it in your mind, or to be "speered at" in regard to an

impending engagement in your family, which you
can neither declare nor deny, — to be told your own
news before it is news, — I wonder if that was not
the devil's fine art in torturing Job? His friends
came to tell him of all these things, which was all
they were left alive for. I think he must have
wished they had not been left alive, and that he
could have found the things out quietly in time for
himself.

This looking over shoulders spiritually into the
page of a life that is barely being written, this pick-
ing pockets of personal experience, is the mean
enormity of which the literal prying into private letters or stealing porte-monnaies are only feeble
types. Yet the social pickpockets run about safely
and respectably, spending their stolen change, and
there is no House of Correction for them.

Arthur Plaice had not got his clothes hung up in
his closet, or his books put up on their shelves, be-

4

fore all that might happen, — well, all that did happen, for what is the use of trying to keep the story
back after a Miss Suprema has seized hold of it? —
was an "I told you so!" in Rintheroote.

There are two ways in which very ordinary men
are influenced by this social force which is brought ·
to bear upon their doings (doings, I mean, which
tend, or may drift, matrimonially), and of which
they usually become aware before the women do.
It either frightens them off, or frightens them on.
Arthur Plaice showed his manhood in that it did
neither with him.

He was probably well aware that all Rintheroote
was peeping and noticing, guessing and prophe-
sying; yet he went in and out just the same, coming
into easy and natural contact with the Whapshare
family, living along precisely as if his life had been
let alone.

Caroline, the pretty one, and the obvious one, of

the Whapshare girls, shielded by this simple " grit,"
as Robert Collyer would call it, of the young doctor,
from the shame and harassment that many a delicate girl does have to go through, — that I have
seen delicate girls suffer from, — of knowing that
a thing has been surmised impertinently, and that he
has heard it, and is shy or cool in consequence, —
Caroline Whapshare went on innocently and quietly,
and kept her little school upstairs.

There was nothing said about the school before?
No; because we came in at the Whapshares' out of
school-hours, at dinner-time, when the pea-soup was
burning; and in the afternoons the little children did
not come.

Caroline Whapshare had not served an apprentice-
ship to any system. She had never been inside a
kindergarten; but she had a garden for little children in her heart, as every woman has who is born
with the genius of motherhood in her, — a place

full of blessed waiting growths and living images
of truth, vital and simple with the child-instinct in
them, — that has never died out of her, but flowers
forth in its heavenly use when the children come, as
it was ordained.

She was full of little, bright teaching thoughts.
Things came to her in clear, happy, object-fashion.
She delighted to tell them again to little, growing
souls, or even to think how she might do it. She
felt always, going through the pleasant mind-and-
spirit places, just as she did once in riding through
a beautiful country, full of farm cheeriness and
woodland beauty, and, far away, unhaunted nooks
and seclusions, "Oh, what lovely places to be a
little child in!"

So she brought out of all her school knowledge
and her later readings, fresh, charming applications.
There was nothing old and trite with her; nothing
that only letters and syllables stood for. The ob-

ject, the very thing itself taught of, was palpable to her imagination; and she made it palpable to the child, in words quick from the live sense in herself, or in some quaint, clever, bewitching little improvised play. She kept a kindergarten without knowing it, or setting it up to be such.

Martha could not keep school; she should not have the patience, she said. She did the Martha-work, and was cumbered, and sometimes cross, poor girl! with much serving.

There were times in that square upper south chamber, where the sun came in on the bare floor, and where three benches and three little rows of desks formed three sides of a quadrangle, and the fireplace was the fourth, with the teacher's table in the corner between it and the window, — times that those little souls will never forget for their early blessedness; times of reciting that were like play, and play-times that were like — oh! what were they

like? when they went "round the barberry-bush,"
or "hunted the squirrel through the wood, and lost
him and found him;" or sang "Chickany, chickany,
craney crow," and ran from the fox that was after
the brood of them. Why, those four plain walls,
and that bare floor, and the three little low benches
that they jumped over for safety, were to them all
wild and beautiful nature, full of fables and fairy
tales that they were playing out. And Caroline
Whapshare was just as young and as pleased, and as
full of "make-believe" and "certain-true" as any of
them.

I think it was the little school, as much as anything, that Arthur Plaice fell in love with.

All winter long the little feet, trudging up and
down the long back stairs, and the little voices,
shrill and sweet and happy, sounded into his heart,
and told tales there; and all winter long the sight
of Caroline Whapshare's face, fair and sunshiny,

grew to be to him a daily bread of blessing that his life had waited for.

He did spend many an evening in the cosey home room, where they were "having the good of their best things;" he helped Charlotte with her sums, and he mended Miles's skates; he went off skating with them all, boys and girls, up the shining river, in the still, keen moonlight; he brought home nuts sometimes, and cracked and picked them, and Martha made pan-candy; he read aloud lovely stories, and books of curious fact, while the sewing-baskets were out and the needles were busy; he showed John how to carve brackets and boxes; he played for them upon the organ; and, on Sunday evenings, they all sang together glorious and tender hymns, or listened while he drew forth from the stops and keys the grand, beautiful meanings of Handel and Beethoven.

He brought into the house a wealth of resource

and companionship; and in return he received —
home. He had not had a home before for fifteen
years; there had only been for him school and
college, and the world.

Why could not people let them all alone, to take
what God was giving, and to make their simple
history?

All the while, the vulgar, hurrying gossip was going about, robbing the sweet, unconscious time that
lives have a right to before they find out their own
whole secrets; interfering, concluding, spoiling.
For while Caroline knew nothing of it, because they
guarded her so, and because she had that kind of
dignity that silly impertinence could never approach
directly, Arthur Plaice and her mother each came
to know it separately quite well; and each felt at
last uncomfortably responsible.

Dr. Plaice was not scared nor small about it. He
had no little pitiful, provoked corner in his mind,

ever so far back, in which he visited upon Caroline
Whapshare the annoyance he certainly did feel.
Her face was just as dear and sunshiny to him as
ever; and he let her see just as plainly the re-
flected shine in his. But he knew that he had a
long waiting-time before him in his life; and he had
a conscience; these two things made a difference.

He began to be busy in his office, or to be called
away now and then, more frequently than he had used.
Mrs. Whapshare had ripping, untidy, or bulky
work upstairs sometimes, and carried off the large
kerosene lamp from below to do it by; and where
mother was, there was always the household. Even
Miss Suprema could see that they were not always
now "lit up and waiting" in the curtained room.
Lydia had a candle, and practised all alone, often;
that was dull. It was all duller than it had been;
they hardly knew when it began to change, but the

winter grew a great deal wearier toward the
end.

It made no difference; they could not defend
themselves; gossip would have something. Dr.
Plaice was " cooling off" now; the Whapshares had
" taken hold rather too strong; " "all the time never
held out; " " it would do Dr. Plaice more good, as a
young physician, to go about and become acquainted
generally." " And what could it amount to?
Neither of them had anything." " It was strange a
woman of Mrs. Whapshare's experience hadn't had
more judgment."

Some of these things crept round at last to
Martha's knowledge. They made her harder and
sharper than ever. She said nothing about them;
but she was brusque, even rude, now and then, to
Arthur Plaice; she was abrupt with her mother,
and with Caroline she was like a thorn-hedge, bris-

tling and thrusting sharp points at her continually,
by way of sheltering her in.

Yet, as Suprema Sharp herself had said, he was
"there right in the house; and there were always
hammers and things." Some pleasant hours were
natural, inevitable; he could not always be denying
himself; neither could even Martha be always on
guard against what there might be no real danger
of, and at any rate was nobody's business.

The days lengthened, and the spring came round.
Mrs. Whapshare had taken Rufus Abell's advice,
and, instead of selling her garden lot, had given
him a two-years' mortgage upon the whole place,
for which he had lent her the sixteen hundred dollars. At the end of that time, he told her, if things
were not easier for her somehow, she could sell at
an advanced value, pay up the mortgage, and have
something left. Meanwhile, as Mrs. Whapshare
said, the children would have two years more of
breathing-time before she walled them in.

III.

HOW THE COMET TOLD TALES, AND SET THE SOLAR SYSTEM IN COMMOTION.

THE houses on the east, or rather south-east side of Ford Street opened by their front and back doors into two different worlds, as the lives of men also do.

One way, there was the dusty, glaring high-road, with the street-cars running up to the corner; the bank, the post-office, the shops, the town-pump, and the hay-scales, all in sight, and constituting what New England people call "the prospect."

The other way, there was green grass, a sloping bank, the shade of trees and wild shrubs, secret stillness and beauty; and the broad, slow river widened out above the dams.

Nobody would have thought it, going by along the front. Nobody would have thought, that behind the commonplace village, with its houses crowding right on to the thoroughfare, was this escape into a hidden and wonderful delight. People did not remember it, although they knew, who lived on the other side, and had close back-yards, stopped short by the yards of Chaffer Street.

The little children knew. Little children always know.

Half Caroline Whapshare's teaching was done, in pleasant weather, out on the "back slope." There was a real barberry-bush to run around.; there were beautiful hiding-places for the chickens, and sly corners for the fox. Above all, there was room for the planets. -

Dr. Plaice came through the long hall of the old house, one day in May, drawn by the open-air chatter of little voices like loosened brooks. He stood there a minute or two in the end door, looking on at a wonderful game, — no less than the game of the stars in their courses.

The roundabout, which dried the clothes on Monday, had its long arms taken out, and piled away beside the fence. To the swivel at the top of its centre-post were fastened stout twine strings, longer and shorter; and each of these was held at its farther end by a little scholar, who, drawing by its tether to a greater or less distance, and keeping the line taut, was joyously revolving in a prescribed orbit, to the time of a tune which Caroline, seated on a low stool at the centre, and personating the Sun, sang to them as the music of the spheres.

Little golden-haired Mercury — the youngest pet pupil, Robie Lewiston — trotted around close by her feet; occulted now and then against her lap when he grew tired. A pretty eight-years-old Venus, sunny-eyed and ringleted, came next; and

then sober, clear-faced, pleasant Ruth Fellman, for
Earth. Mars was a sturdy, rollicking, rather un-
manageable fellow; Jupiter, Saturn, Uranus, were
the big scholars, in the edge of their teens. Farther
into space Caroline did not try to go; nor could she,
without getting into the river. It was enough for
all practical purposes.

By and by (this was the best part of the play)
Caroline lifted up her hand, and forth started a
comet from behind a gooseberry-bush. From away
down by the bank of the river he came, describing
his parabola among the planets, bearing down toward the Sun, crossing orbit after orbit, but never
when the heavenly body was there. This was the
"steering." It was as great fun as coasting down
hill among multitudinous sleds. He took his sight
from the start, and threaded his way, bobbing under
the lines, and, wheeling at length close around with
little Mercury, shot off again upon the other side.

Dimmy Pickett did it; a pennon of white muslin, fastened around his head, flew behind him. This was the comet's tail. Dimmy was only seven years old, little and bright. A larger, duller boy could not have done it.

When the play was over, the planets, out of breath, came up around the Sun; and the Sun asked them questions.

" What are the strings meant for? "

" Gravitation, that ties them to the sun."

" What is your pulling away as far as you can for? "

" Centrifugal force, that makes them fly off."

" What do both together do? "

" Keep them going round and round just in their own separate places."

" Are there really strings up in the sky? " asked little Venus.

Caroline held up her finger, and beckoned to Venus. Venus came.

"Why did you come to me? I did not pull you with the string."

"You beckoned."

"God beckons."

All the little planets were still. There was silence in their heaven for the space of half a minute.

Then Dimmy Picket spoke.

"Suppose she had had her back turned?"

"Every little atom in the whole world of worlds has its face toward God."

"What do they pull away for, then?"

"God gives them a will of their own, to go a little way of their own; but they cannot get beyond his will. The two wills make the beautiful glad motions, and all the life and the glory.

"There are anemones down by the spring. Who

5

will come this afternoon, and go with me to gather them? "

Caroline had given them their bit of physics and metaphysics. It was enough for this time.

Everybody would go and gather anemones, — everybody but big Jupiter. He did not say anything; he wanted to play football.

"May I go too?" asked Dr. Plaice, coming over from the door.

Caroline had sat with her back toward him. She started a little, and flushed.

"It is the children's walk. Will you have Dr. Plaice go too?" she asked them.

" He doesn't belong," whispered Venus, shyly.

" Oh! I'm the new planet, — the far, far-away one, that only comes in sight once in — ever so long. I've been a good while getting here. But I'm discovered now, and must be counted in. I belong; truly I do."

Something made the pretty Sun change color yet more at this. Among them all, nobody had the presence of mind to say him nay. So the doctor said he would come, and bring his microscope with him. After the tremendousness of things in general, they might like to descend to something small and particular.

Dimmy Pickett stood staring, in a queer, bright, eager way, while the plan was settled. He looked at the doctor and at Caroline, as if he were making a bewildering computation, astronomical or otherwise, too large for his small head.

Caroline did not notice; she was busy with little Mercury. But the doctor saw it, and had an end-of-the-world instinct that the comet was bearing down upon him.

All at once, the erratic little luminary did bear down upon the Sun, displacing Mercury.

"See here!" said he, breaking out with a shy

bravado in a child's loud whisper. " I know something, Miss Caroline, — I do ; only Flipper told me not to tell."

" Then," said innocent Caroline, " be sure you don't. You won't ever be a man, — a splendid, honorable man, — if you tell things that you ought not. And say ' Philippa.' Your sister has a pretty name ; but ' Flipper ' isn't pretty."

" Everybody calls her ' Flipper.' She is ' Flipper ' ! " returned the comet, half inclined to be a little sulky. He had expected to have his secret teased out of him.

Dr. Plaice caught the last sentences as he turned away quickly, for fear of what might come next. He walked back into his office with an excited perplexity in his mind.

How long could he save Caroline from this ? And what ought he to do ? Go away ? or stay, and do

that which he had hardly made up his mind would
be right to do?

He sat down in his corner chair, near which the
little passage and the blinded east door were open,
letting in the soft summer of a few hours that the
May day was giving.

He had hardly sat there two minutes, when
little steps came by around the corner, and little
heavenly bodies — three or four — made a constel-
lation just outside the folded blinds.

He could see them as they stood. The Comet
looked big and red and portentous; little Venus
was sparkling and coaxing.

"Tell me, Dimmy; just me, you know."

And Earth and Jupiter crowded up close also to
hear.

"I s'pose Flipper meant not to tell *her;* besides,
she's always telling everybody not to tell everything. And they do. She does."

"Grown-up people tell the most, I think," said Venus, gravely. "They keep all the telling and all the cake, and say it isn't good for children. Is it about us, Dimmy?"

"I told you 'twas. By least it would be some time. She said it would be a forever vexation."

"Vacation, you mean, Dimmy," said elder Earth.

"I say vexation at home; and Flipper says it is vexation. So now," said Dimmy.

"I shouldn't like a forever vacation," said Ruth Fellman, waiving the point.

"But it would be," persisted Dimmy, "if she went and got married. And Dr. Plaice is her beau. Flipper said so."

"Poh!" said big Jupiter, and walked off.

Earth and Venus looked at each other with a wide wonder in their eyes, and set their little white

teeth suddenly very tight upon their under lips.
It was a tremendous secret!

Venus came to first.

"Well, it must be pretty nice to have a beau,"
she said.

"Mr. Dimmy Comet!" said a voice behind
them. The blind opened, and the doctor stood
there.

"Allow me to beg the honor of a further ac-
quaintance with so well-informed a gentleman.
You will please to walk into my office here."

Dr. Plaice's hand was on Dimmy's shoulder.

"Oh, my gracious!" cried Earth and Venus
simultaneously, and simultaneously rushed down a
broad vista of space, that is, the village street, that
turned between the tin-shop and the tailor's.

That light hand on Dimmy's shoulder was not to
be mistaken. He walked in up the step as a little

boy does walk in when his sins have found him out.

Dr. Plaice closed the door.

"Take a seat, Mr. Comet," he said politely. "The arm-chair, if you please."

If he had put him on a cricket, or let him stand, it would not have been half so bad. The arm-chair was high, formidable, and awfully suggestive. The tone of the "if you please" was unrelenting. The doctor might be going to pull all his teeth out; but he was without remedy.

Dimmy hitched up backwards into the great chair, putting his heel upon the forward rung, and hoisting himself by the arm. Seated there, his legs hung ridiculously short and small.

"The leading object of my life," said the terrible doctor, turning to the mantel, and taking up his meerschaum, "is enlightenment. You have enlightened me very much indeed within the last five

minutes, Mr. Comet. I feel exceedingly obliged
to you, — and to Flipper." And the doctor filled
leisurely the bowl of his pipe, pressing the tobacco
down evenly.

"Smoke, Mr. Comet? No, I thought not.
Judging professionally, I should say that your con-
stitution was not quite — up to it."

Dr. Plaice struck a match, held it to the pipe,
and took a whiff or two, then drew a chair, and sat
down himself.

This was awful! How long was it to go on?
How long did it take the doctor to smoke his pipe?
Would he keep him there all day mocking at him?
Would he ever let him go? And what would
Flipper say?

Dimmy twisted his short legs desperately, and
untwisted them hazardously, and recklessly twisted
them again. He squeezed the rim of his little soft
felt hat into a great many doubles, to correspond

with his legs; then he let it out, and squeezed it
up again. He began to grow alarmingly red and
swelled in the face with mingled shame and fear
and indignation.

"Your ·news was very interesting, Mr. Comet,"
resumed the doctor; "especially to myself. For
that reason, and for another that I will mention
presently, I should prefer that it should not be
spoken of in like manner again. Do you understand?"

For all answer, Dimmy struggled with his legs
again, and obliterated his cap.

"The second reason is, that it does not happen to
be true. If it were, I should be likely to tell of it
myself. A gentleman, Mr. Comet, does not speak
of other people's personal affairs until he is author-
ized; and he never repeats things that he hears, in a
whisper, with a 'Don't tell!' neither, I think, does

a lady. In the first place, ladies and gentlemen do not very often hear those things at all."

Dimmy's redness grew ominous. He winked very hard. These were very grown-up words of the doctors; but instinct .translated them. He learned a half page of dictionary at least, in these five minutes, that he never forgot. He was very much ashamed, and he was very mad. ; His legs were in such a snarl with the chair by this time that it was hard to tell which was human and which was mahogany; his face was big with tears that he would not cry, and his hat was pretty nearly hopeless.

At last, two words came forth, very much thickened and swollen themselves with their long restraint : —

" By George ! "

Dimmy lisped a little on his g's; and the expletive sounded like something huge and soft, flung

with great force, and hitting as hard as it could.
Dr. Plaice laughed out; he could not help it; but
then he immediately got up, and came over toward
Dimmy, with his hand held out. He did not wish
to humiliate and enrage him utterly. He meant to
treat him really like a man at last.

"That is all, Dimmy. Now let's shake hands,
and be friends. You don't like being talked to like
a mean little man? Well, you can wake up from
that bad dream all safe at seven years old, with
twice your age yet to grow in, and to make what
kind of a man you will. Miss Caroline told you;
if you want to be a 'splendid, honorable' one, don't
do any small meddling things, or tell any small,
meddling tales."

And Dr. Plaice kept hold of Dimmy's hand till
his legs untwisted, and he was slid safely down out
of the big chair. Then Dimmy put on his cap,
pulled it very much over his eyes, and departed

meekly and swiftly. When he was around the
corner, however, behind the tin-shop, he paused,
pushed his cap up into its place, took a good long
breath, and said "By George!" again. But there
were things in this "By George!" that had not
been in the other. Out of it came a good deal
in the boy's life that would not else have been
there, and that we shall not follow him on to tell
about.

The first resultant was his going with the walk-
ing-party that afternoon, in spite of the tingle with
which he thought of it; which, if he had not been in
a pretty fair sense a "by-George" character, one
would hardly have expected him to do. He had
two minds about it; but the spirit that swore by
the king that was in him prevailed. He wouldn't
sneak off, afraid. He would face the doctor and
those girls. Besides, he would stop the tattle; that

is, he thought he would.   There was a good deal of
the royal in this for seven years old.

Venus was in the middle of a knot of girls when
Dimmy came upon the field.   He watched and loi-
tered, until she emerged for a minute, and he caught
her upon the edge.   Then he sauntered by, close to
her, his hands in his pockets.

"I say," he said low, over his shoulder, "don't
tell of that, you know.   'Taint true."

"My sakes!" cried little Venus, coming quite
away, and going on with him; "I have told."

"Poh!" exclaimed Dimmy, in disgust.   "Who?'

"Just Aurora, my best friend, you know."

Now, Aurora was just the biggest little chatter-
box in the whole school.

Poor Dimmy began to find out, to his dismay,
how hard it is to catch up with a mistake.
He thought of Jupiter, too, off in his bigger orbit,
with the village fellows.   What might not he say, in

his big-boy fashion, worst of all, notwithstanding his "Poh"? The little Comet was very uncomfortable, and wished with all his heart, that he had kept his tale to himself.

Aurora was nudging and whispering, walking behind the doctor and Miss Caroline, with her other best friend, a larger girl, Laura Frances. It was plain there was no knowing what might come of it. The whole solar system would have hold of it, and what a blaze and whirl that would be!

Dimmy marched up to Dr. Plaice, at his open office-door, when they were back again, and the girls had gone.

"I can't help it, after all," he said, without any antecedent to the "it." "I tried to stop it, and it won't."

"It isn't easy to stop a thing that is once started. There's a law of nature against it. But I'll see

what I can do, Dimmy; and it is all right between
you and me, any way."

Dimmy's throat felt queer; and he came very near
saying "By George!" again.

The sun was going down, and the air was just as
sweet and tender as it had been all the day. Windows and doors stood wide, gathering in the rich
feeling of June from the May air. Dr. Plaice came
round through the hall again.

"Miss Caroline," he said, "the Golden Gate is
open. Will you go down and see?"

The Golden Gate was the opening up the river
where the west shone in, and filled up all the water
aisle with a mist of glory. Far and deep between
the trees that closed on either side lay the burning
splendor whence the tide flowed down; and violet
or crimson bars would lie across as the flame faded,
or flecks and burnished lines of yet intenser fire be
thrown up like isles and coasts along a dazzling sea,

and all gathered, as it were, into one focus of light, for the wooded fringe and the high banks of the stream covered at right and left the stretch of the horizon, and left all heaven to be imagined from its single unclosed door.

So they went down to the river-side. The sloping bank shut out house and street and all [the village sounds. Office and school-room, and all the ways by which their living and everybody's else went on, were behind them. Nothing was here but God's beautiful world that his souls are born into, and before them the Golden Gate lay open.

"It is like a beautiful secret," said Arthur Plaice.

"It is like the heaven inside and behind," said Caroline, softly.

"Yes; it is like that. It is that heaven *is* the great, beautiful secret. There is a piece of it, Caroline, that I have wished to tell you. Only the other side, there is still the dusty street."

6

Caroline stood utterly still.

"I am afraid I have no right; because — " his pause became a period. "I have earned just fifty dollars all this last year beyond what absolutely had to keep me," he said, speaking it out quickly. "Your little school is better than that; and so I have no right to tell you beautiful secrets by the river-side, and then lead you out into the toil and dust."

"You mean that you have been paid just fifty dollars," said Caroline, looking at him very proudly, and then turning away again; "and — I don't care for the dusty street."

"And you do care — ?" asked Arthur, eagerly, bending down to look after the shy face.

Caroline flushed up like the sunrise that tells God's morning story without any words.

Arthur Plaice felt the joy of his morning; but he was a man, and wanted speech, — just a word, ever

so shy, ever so small. He forgot his own un-
finished speaking.

" Translate," he whispered.

"I do care," said Caroline, quaintly and tremu-
lously, " for the beautiful secret — which you didn't
tell me."

And then the secret was told.

" I think they have gone *through* the Golden
Gate," said Lydia, turning round from her organ,
when she could no longer see her notes.

" I believe so too," said the mother, seeing them
come up the old stone step at the end door ; but she
said it to herself.

She stepped out from the little dining-room where
the tea was ready, — split-cake, toast, and a pink
square of delicately broiled smoked salmon, — and
met them in the dusk of the long, old hall.

"Will you come in?" she said to Dr. Plaice. "We are just ready."

"I will come if you will let me, — mother!"

He had got her hand fast with Caroline's in his own, as he said it.

"O you two children!" Mrs. Whapshare answered, when she had got over a little sob. "How long you have got to wait!"

"We can't help that," said Arthur. "It won't be any longer than it was before. And we should have waited. I suppose we have been waiting, ever since we both were born."

Dr. Plaice took care to meet Dimmy Pickett the next morning.

"I've stopped it, Dimmy," said he, holding out his hand.

"How?" said Dimmy, explosively.

"As the Indians stop the fire from chasing them on the prairies, — kindled it at my own end. I

want your congratulations, Dimmy. I am engaged
to be married — some time — to Miss Caroline Whapshare."

Dimmy drew back his hand to pull his hat down
over his eyes. He shuffled with one foot back and
forth upon the ground. He was overwhelmed by
this real, grown-up news, told him with his hand in
his friend's just as if he had been big enough. He
did not know what to do with it, or how to get
away and leave it. All at once he pushed his hat
back again, stood square upon his feet, and looked
up.

"Are you making fun of me now, Dr. Plaice?"

"No, indeed. I am telling you my good news as
my particular friend, whom I told yesterday that it
wasn't true. You'll wish me joy, won't you?"

"Yes," said Dimmy. "But, if you want anybody
else to know it now, I guess you'll have to tell 'em
yourself. There's Miss Suprema coming."

And Dimmy vanished round the corner, and into the school-room door.

Dr. Plaice stood still and laughed. " That's the brightest boy in Rintheroote," said he to himself.

Miss Suprema came up.

" Why, doctor, what is it? What have you done to Dimmy Pickett? "

" Told him some news, and got his advice. The advice, I think, was excellent; and I am sure my news was."

Then he told her the news; and she forgot to ask him anything about the advice.

When he went back into his office, he saw her, through the blinds, standing in one of her awful equilibriums. Whether she should keep on, down the village-street, taking her chances as she went, or turn about, and go straight up to Mrs. Benny Dutell's, before she heard of it from anybody else? She could not expect to be first with everybody;

she must be first with Mrs. Dutell. So the great whirl within her set her off in a right line at last, and she went up the street like a cyclone.

The doctor drew up his shoulders with a laughing shake, turned to his desk, and sat down.

Sat down to his desk and his books; and knew that he began, that moment, the days of a hard, uncertain waiting. The news was told; the fire had run; he had made a safe place to stand in; and now he must only — stand. That makes a long chapter; the Apostle Paul knew that, but it is not a chapter for a magazine.

"It is all there can be about it for ever so long, Arthur," Caroline herself had said to him, in the first, blessed, sober, certain "talking-over."

"Mother could not do without me, and my little school, until Lydia is ready with her music, and John gets some sort of salary that will more than pay for his tickets in the cars and his lunches in the city. I must stay by home, you see. I shouldn't be worth taking away if I wouldn't."

For two years there was no new point reached in this, their story; none but the little shining points that count in "the kingdom;" in the inside beauty that lies away from the dusty street; that holds all the loveliest secrets, and the least of them sometimes the loveliest; and where the Father that seeth in secret keeps his own inner blessedness hidden fast with the hearts of his children.

But in two years the outward may halt step with the inward till the hobble grows wearisome and painful.

In two years Dr. Plaice had only put into the bank two hundred and fifty dollars more. In two years Mrs. Whapshare's face had gathered new lines, and Caroline's had grown a little thin and pale with the constant pull of school.

Martha was two years crustier, and more like an

old maid, while her service in the household was
more comprehensive and invaluable than ever.
Lydia and John were growing up to the realization
of the hard tug of life, and the knowledge of the
many wants and wishes that must go unmet.

Suprema Sharpe had had two years in which to
find herself often at default for fresh aliment of
news, and driven to turn and worry and recrunch
the old; as a dog keeps a bone buried, and digs it
up once in a while to try for a little more marrow
in it.

Every now and then she dug up the Plaice-Whap-
share bone; and every time she set it forth in sorrier
fashion, and yet "bonier" light.

"The doctor was tired of his bargain; he hadn't
much the look of a satisfied man; if it was ever
coming to anything, why didn't it come? The
Whapshares held on well; she would say that for
them."

Or, it was " a shame for Mrs. Whapshare to keep
Caroline toiling on at her school for her. Why
couldn't she marry, and keep school to help herself? Car was growing old; she had got gray hairs
on her temples. No doubt they were awful poor;
everybody knew the place was mortgaged; and old
Rufus Abell didn't lend his money just to get it back
again. There was Lydia flourishing away on that
organ. Much she'd ever make of it! She'd better
have been running a sewing-machine."

In two years, Zerub Throop was dead, and nobody could find out, for a good while, what he had
done with his money. By and by it came out that
there was a will, and that Rufus Abell was executor.
Of course; Rufus Abell executed everything.

Mrs. Whapshare took to having little, nervous
starts every time Rufus Abell came round the corner.
She could not shake off the notion that news was
coming to her yet, from old Zerub; from old Zerub —

and the Lord; for she remembered always that about the king's heart; and she knew that in the inward light of things she had a right, and that the Lord and his angels live and work continually in the inward light, where man can neither see nor reach.

But Rufus went and came, and never stopped, or even looked up at the Whapshare windows. It was plain that he had no thought of any contingency for them.

All that was known about the will was, that it was an odd one; as it would not have been Zerub Throop's if it were not. That nothing was to be settled — save certain legacies, the chief of which was to Sarah Hand, providing for her and for the cat — for five years; only the property to be taken care of, rents and dividends collected, and all to wait that time, for any claim that might arise; failing which, it was then to be devoted to certain specified public uses.

Rintheroote was exercised to conjecture what that possible claim might be. A secret marriage, — a child, — half-a-dozen children, perhaps, adrift somewhere, liable to turn up?

Rufus Abell held his peace; indeed, he had nothing else to hold; the will registered, and open to any reading, only said just that: "For any claim upon said estate that may legally and within that time arise."

But Rufus Abell did call one day. The mortgage-debt was falling due, and the garden-lot would have to be sold.

This was how it was with the Whapshares at the time the queer thing happened which nobody will believe, and which Mrs. Eylett Bright will tell of in the next chapter.

IV.

HOW THE GHOST MANAGED. — MRS. EYLETT BRIGHT'S STORY.

My dear, I well tell you all about it. It was a haunted house. It was all explained by simple causes, — yes; but it was a haunted house, nevertheless. It is a haunted world we live in, for that matter, Dora Dutton.

You see there are so many of us, — so many little Eylett Brights; I like to call them by their whole patronymic, it suits them so well, Dutton, dear.

We all needed the country that summer. I was run down with change of servants, and nursing; little Thode had just crept out of scarlet fever,

6

with the tattered shreds of his dear little mortality about him, wanting all sorts of patching up; and the other children had had it too, more or less; mostly less, thank the good Providence! We all needed the country, — doctor said we must have it; but there was Eylett tied down to his desk, and the two thousand wasn't any bigger for us this year than ever before.

The country is so wide and free; and yet it is so hard to get a place in it, — a place for ever so many little Eylett Brights!

We wanted a large house, and we wanted it furnished; there must be plenty of out-of-doors, and yet we did not want a "place" that would have to be kept up. People who were going to Europe, and had out-of-town residences to leave, must leave them to their own sort, you know; carriage and lawn and garden people, who would have gardeners and grooms. It was as much as ever we could do to have Onie and Ann. More; for they were both going to leave. They had objections to the country. So we got Margaret and Ellen from the intelligence office, — the same article, you know, with a new label; and there isn't much variety in the labels, either. It is wonderful how we have rung over the changes, — Margaret and Katy and Ann; Bridget and Ann and Katy; Bridget and Margaret and Ellen; and how natural and of course the name sounds, whichever it is, when they tell it; and how the impression of the whole successive multitude drifts and runs together in our minds into the image of one great, awful, representative, — kitchen creature !

Well, we searched the papers, and we searched the country; we had spent fifteen dollars before we knew it, running out and in to see things, and conclude they wouldn't do. So we kept quiet a while, and almost gave it up. Eylett said we might hit upon something by and by, when somebody's house was left on their hands, too late for a high rent or a whole season. I didn't see how, though. I told him it would have to come and hit upon us; we couldn't afford to go after it any more.

Things do come and hit you if you only stand still because you must, — not because you're lazy.

One day, at the counting-room, Mr. Haughton was asking Eylett after his family. Eylett told him he was getting along; but they needed a change, and it was not easy to make a plan that would suit in all ways.

"Take a house a little way out of town," said Mr. Haughton.

"I've been trying to," said Eylett, "but the house I want doesn't seem to be anywhere."

One of the boys came in from the bank just then, and heard it.

"I know of a house, Mr. Bright," he said; "but it's rather a queer one, up over the hill, out of our village; and to let cheap, I guess, — old Zerub Throop's. He's dead, and things aren't to be touched for five years. But the house can be hired just as it is, if anybody likes. It is a jolly big one, and an old garden and fields all round it. Why don't you come out and see it?"

Eylett guessed he would.

And so, one day, we went out to Rintheroote.

Why, you see it was splendid! All that great hill, and the sunrise on one side, and the sunset the other! But, as to the house, it seemed as if the day had always had to climb over and round it, and had never shone through it. Such a musty, shady, lo-from-the-tombs old place you never got into! The front door was all grown up with weeds and vines. It was tall and narrow, with an old-fashioned fan-light over it. It looked as if

nothing had ever gone in and out but coffins, I told
Eylett.

We found a woman in the village who had kept
house there; and she went up with us, and showed
it.

"It's in good order," she said; "the front part's
clean, because it aint never been dirtied; and the
back part's clean, because I done the scrubbin'."

There was one real lovely room across the ell,
upstairs, at the end. Four windows, — east,
south, and west, — the sun and the soft wind just
rioting through.

"O Eylett!" I cried, standing in the middle,
"Here's the summer time and the beauty! Here's
the life of the house!"

"Yes'm," said Mrs. Hand, "here's where 'twas.
But I'll tell you what, 'taint more'n fair to let you
know. I don't believe it's all gone out of it. *I -
don't believe, in my soul, Zerub's done with it!*"

She spoke in a hushed way, as if there might be some one listening.

"Done with it? He's dead!"

"Yes'm; that's just why you can't tell. I stayed here a month afterwards, and I had — well, experiences. If I was you, I'd shet it up."

"Shut it up! I shall put the children into it."

"That may do. Maybe he'll quit, then."

I had my doubts about that conclusion, if I hadn't about the ghost. I couldn't think, if he wanted to come at all, that old Zerub, or any other rational spirit, would come back the less for, — you needn't laugh, Dutton; I don't care if they are mine!"

"See here, my good woman!" says Eylett turning round sharp, "I can't come here if my servants and children are to get hold of this nonsense. Has it been talked round in the village?"

"Not from me; I've held my tongue too long for Zerub to begin chattering now. I always left all

his affairs to hisself, an' I do yit. But this is your
affair, kinder, if you're comin'. I jest eased my
mind.".

"It shall be the play-room, — the day-nursery,"
I repeated, ignoring the nonsense once and forever.
"And here," said I, going back into a small adjoin-
ing chamber, "I'll have my sewing-machine and my
writing-desk, and all my little things and doings
that I want close by the children, but not mixed
up and crowded with them. We can be grand
here, Eylett. There is no end of room. As to
those front parlors and bedrooms, we'll fasten back
every blind, and fling up every window, and let
June do the rest. We'll come, Eylett, won't we?"
I concluded after my wife — fashion, — a decision
first, and a question afterward.

So we went down into Rintheroote, and found
Mr. Rufus Abell, the agent; and Eylett put in the
ghost story in the way of business, and got off fifty

dollars for that; though I told him men always came
out with the very thing they didn't want mentioned; and we took the house for three hundred
and fifty dollars, and could stay the season, — three
months, or six, as we had a mind.

But we were not to ask to have the first thing
done for us, and we were to alter nothing ourselves.
These were the conditions.

We had a splendid time moving. You know I
don't mind trouble; and the children were as gay
as larks. We didn't have much to move, either;
only our clothes, and the few things we couldn't
live without, and to send the rest right off to a
store-room; for we gave up our house in town, of
course.

Margaret and Ellen, gave warning the second
morning after we got there; that we expected. All
we hoped for from them was to get through the
flitting; though how they could, with the sun shin-

ing as it did, and the clover smelling, and the birds singing, I don't see. I should as soon have given warning in heaven, — as, to be sure, I suppose some folks will!

Well, we didn't care; it was all fun, nobody was going to call, I could just put on a calico wrapper, — keep it on, I mean, — and take right hold, if it came to that; and we set Mrs. Hand to inquiring for us in the village. In result of which, after three days of the "warning," and three days more of the "week" that they wouldn't stay, and hardly ever will, and you hardly ever care to have them, since the days of warning are in themselves so like the days of doom; and after yet three other days of expectation and hard work, and baker's bread, there came to "our ha' door," and when that was opened into the best — I mean the dingiest — parlor, a — well — these presents : —

A hat and feather, — that is, a very remarkable

and exaggerated piece of a bird, that was neither wing, tail, nor breast, but enough of it for all three, attached mysteriously to the middle of a forehead; an emphatic chignon, a very much fluted and hitched-up alpaca overskirt, and a pair of tall-heeled boots, on which all the rest walked in.

What else should have come, unless, indeed, it had happened to be a man? These, you know, are the things which stand for a woman nowadays, and make up the general presentment and expression of her, confounding distinctions; so that the pieces of a woman in the windows of the great furnishing shops, "articulated" on wires, hint out something rather superior, on the whole, to most of the specimens which articulate themselves, and are seen about the streets.

The "articulation," in this instance, announced herself to me, looking at her with a puzzle and a question in my face, as "a girl." An American girl she was too; no Irish, we found out gradually, would apply. Although Sarah Hand had been reticent, Terence Muldoon, who chored, and chopped wood, and "fought and carried" for old Mr. Zerub- babel Throop, and who stayed by to "garrud the hoose," with Mrs. Hand, during the month of her closing-up services and administration, had not been so; and there were vague and terrible rumors afloat in the Irish stratum of society, and the uni- versal Irish mind was set against the "owld Throop place an' its divilments." This came to us by de- grees, as our own experience developed.

"I'm the girl," said the articulation, "that Miss Hand was to look up. She's my Aunt Sarah. I'm a dasher."

"You're a — what?" said I, explosively, in my astonishment.

"A dasher: — A dasher down."

I just stared. I began to think she must be a

lunatic. And a lunatic who announced herself as a dasher down might not be the subject of a form of hallucination one would like to have illustrated in one's parlor.

But, while I stared, she added mildly, "That's my name."

"Oh!" said I, relieved, and catching my breath. "Just spell it, if you please."

"A,d,a,s,h,a, — Adasha; D,o,w,n,e, — Downe; Adasha Downe."

"Thank you. It sounds rather terrific, you see, before one knows, especially for a person who is to handle cups and saucers."

Adasha gave a bright look out of her eyes without moving a muscle of her very round, and very large, and very solid face.

"There's many a one gets a name, you know, for a thing they never did." Then she smiled widely. She could not help it; she must do it widely, if she

smiled at all. It took very little exertion, and but
slight play of her lips; for her lips were ample, and
behind them were white teeth that needed generous
accommodation.

I liked the smile and the bright look. I began
to think of engaging her; up to that moment I had
only thought how to get rid of her. I asked her
if she could make bread and hop-yeast; if she could
wash and iron; and if she would do anything else
that I might ask of her, and tell her how.

She could and she would.

" Will you take off your things and stay now? "

" Well, ma'am, you see, in my suit and my heeled
boots and my hair, I don't really see how I could.
But I'll get a bag o' clo'es, and come back in half an
hour."

" Very well."

She did.   And so we had Adasha Downe.

That was all we had; and we found it was all we

had to hope for. For love, nor money, nor for Christian charity, we could get no soul to offer or consent. We tried for three weeks; and then we settled down, until the prejudice should wear away, to a plan that we fitted to the case. A boy to do chores, and a woman to come three times a week, and wash and iron and scrub. Then, with all the children, and their summer liberty, on my hands, I thought of another expediency, — a young girl as a sort of governess-companion, who might keep them up in their A, B, C, and their tables, tell them which side of the world they were on, and a few preliminary items of like importance; sew on a string or a button now and then, and help me in such things as I daily put my practical hands to.

We found her; she was foreordained.

Do you remember little Car Whapshare, the youngest girl at dear old Cradley School the last year we were there? She lives right here in Riutheroote; and she had kept school until she hadn't much face left; though what she had kept the pretty in it, as the child's barley-sugar keeps the clear and the sweet down to the last thin needle of identity. She was engaged to marry — in this life or in the life everlasting — a splendid fellow, the young doctor of the place. But the old doctor wouldn't let go, and the old patients wouldn't change; and so he was getting — excellent practice and very limited pay; and Car's mother was poor. And that's the way things were with them; and they couldn't be much more wayward.

Arthur Plaice — her doctor — said she must give up teaching, for all summer at least. She was in a worry. But then, there was I in a worry too, up there on the hill; and the worries of the world do, once in a while, when the right ones are thrown together, turn suddenly, by the beautiful chemistry of things, into a blessed mutual content.

Car Whapshare came to live with us all summer.

And it was just after she came, mind you, that the signs and wonders began.

How we three — Car, Adasha, and I — did work, letting the besieging pleasantness into that old house! Adasha, cleared for action, without her heeled boots and her hair, — that is, with only a reasonable amount that you might believe in, gathered up with a screw and a double behind, and fastened with a rubber-comb, — without any humps or hitchups, — turned suddenly into an individual. That was a refreshment and a confidence. I suppose there is a beauty of "the all," — Emerson says so; but you do want caches; the world will never make up the nicest kind of total by rubbing out its units.

We could not alter; but we could innovate and renovate. We rolled back the heavy worsted damask curtains on their old-fashioned gilded poles,

threw wide the blinds, and let the summer in. We
turned the musty old chairs and sofas out on the
grass; we cut away the thorn-branches, and the
twisted stems of creepers from the choked-up
porch; and we left the high, narrow door open all
day long, so that a column of sunshine poured itself through that way in the morning, and bars of
gold shot slanting across from the windows of the
south parlor through the noontime. When the
house was sweetened full of it, we began to shut
the green blinds again in the midday, and only
leave the air to filter in from over sun-basked fields
and tops of clover.

"We'll drive the ghosts out," I said, gayly.

"They'll be driv' out or stirred up," said Adasha
Downe. "I don't s'pose we can tell which till
we've tried."

Mrs. Hand came up several times to see us.
Partly because of her niece; partly because of the

cat, which was her charge, but which she could not coax away with her; and partly to ask me privately every time, and with solemn emphasis, just before she went away, "if we had noticed anything."

"Nothing," I told her at last, "but that black cat. She haunts the house. There's something awful about her. She steals round everywhere, like an uneasy spirit; but she won't come in and be tame. I have met her in the rooms and on the stairs; but the minute she sees anybody, she's off like a black rocket, with her tail straight up in the air, and as big! The children have found a kitten; they pet that, and the old cat stands away off and watches. She is like a human mother, that lets her child be taken in where she doesn't feel willing or worthy to go. She behaves like a bad conscience."

"Zerub Throop hadn't a bad conscience. He warn't givin', nor he warn't pious; but he was a real righteous pertickeller man."

"I never thought of Mr. Throop, Mrs. Hand. I was speaking of the cat."

"All the same. She's in it. She knows," said Mrs. Hand, impressively.

"Cats are signful creatur's, about weather, an' sickness, an' such; an' they have a feelin' for other-world things, too, you may depend they do. They see in the dark. What does that mean? It just corresponds. Do you know how hard it is to keep a cat out of a dyin' room, or where a corpse is? You just wait and notice."

"Oh, for mercy's sake, don't!" I cried out, almost with a shriek.

The woman was growing ghastly.

"La! I didn't mean anything. Like as not you'll never have a chance. But that's a fact. It's the reason why they stay round places so. Everything isn't gone, and they know it. Why, live folks leaves something of theirselves in the places where they've been and acted. Now, whenever I heerd them noises, that cat was alwers yowlin' alongside. Way off, maybe, or even afterwards; but she always jined in, — or Amenned."

"Mrs. Hand, what were the noises?"

"I don't know. Kind of stirrin's, — soundin's; everywhere to once, distant and down-like, but strugglin' an' risin' up. I can't tell you what they were; but the old house seemed all breathin' alive with 'em, as if they might bust out anywheres. I'll tell you what I think. If ever you hear anything, you'll hear more. It seemed to me as if 'twas only a kind of gettin' ready, a-gropin' out. You wait and notice."

"If only you wouldn't please say that!" cried I, nervously. The words were growing awful to me. And then I laughed at myself for minding them, or any of it, as I bade Mrs. Hand good-morning at that pleasant east side-door, opening out into the warm, living breath and glory of the perfect June day.

Well, the children had their games all day long; their blocks and their baby-house, their tea-parties and their soap-bubbles, in the bright ell-chamber; and they played horse, driving each other with gay, knitted harness and reins, up and down the long passages of the old house; and they went to bed at night in the west rooms, back of ours, where the twilight lingered till they were fast asleep; and I said to myself, "They take up all the time, and they fill the house full; what else — if there were any-thing — could creep in? Their little plays, and their little prayers, and their little dreams, and their sweet sleeping breath, — why, it's a *home* now, brimming over with them. Bad vapors couldn't come up through the fair, full fountain."

And so, after the happy, tired day, I went to

sleep myself, and slept as having angels about me.

There was one thing we had to do to that ell-chamber. We had to take the door down. It was a modern door, put up since Mr. Throop came ; and it lifted off on its hinges. The reason we could not have it on was, that it shut with a horrid spring-lock. We couldn't have the children getting shut in there every day, and having to be taken down outside, you know, with ladders.

Eylett and I had the north-west front bedroom. There were two large rooms, and a little one tucked in between, on each side the hall in the main house ; then the long ell ran back, and there were three or four in that, besides the attics. Caroline Whap-share slept in the large one back, on the south-east side, and the children, as I said, were in the rooms behind ours. Nobody slept in the ell. Adasha

Downe had the little room next to Miss Whap-share's.

Somehow, in the great rambling place, we did like to keep all together at night. There would be thunder-showers, and there might be burglars; nobody believed in anything else or farther off. The children never heard a word. I found I could really trust Adasha Downe.

Whether it was the fatigue that gave us such sound nights, or whether there never was anything to wake us up until the night I am going to tell of, I don't know; but so it was, that, for a week or two after my talk with Sarah Hand, we might have been the builders and first dwellers at Throop Hill, for all sign we had from the " soul of things " in its old timbers or out from its far corners.

Then, all at once, something happened.

I had gone to bed one evening at ten, and had

had my first two hours' nap. Suddenly I sat up, wide awake.

Something crashed me awake; a great resounding came with me out of my dream; and I listened mentally in as great an outward silence, to hear what it had been like.

A ringing, clattering, metallic sound, as if a tin-man's cart had been upset outside, or a great sheet of thin iron been shaken or struck upon somewhere in the house.

Had I heard it? or was it only that all my nerves had suddenly vibrated with some tingling shock, and waked me with a feeling of such sound? It was "all over, everywhere," as Mrs. Hand had expressed it; either all over me, or — creation perhaps.

Why did not everybody in the house wake up?

While I held my breath and wondered, it came again. Now I knew that I heard it with my bodily

ears. But what I heard, I could neither conceive
nor tell.

"My gracious, Eylett! what was that noise?"

I had my hand tight upon my husband's shoulder.
But Eylett was lying on his right side; and he
could not hear with his left ear.

"Noise? I don't hear any. Let me move. Let
me get my good ear up. What was it like?"

"I don't know. Like a ringing, or scraping, a
rattling, a reverberating, crashing and hollow, far
off and all round. In the air. As if the house was
a Chinese gong, and somebody was walking in the
middle of it."

"All that? Pooh! You've been dreaming."

"No, I haven't. I've been sitting straight up
with my eyes open."

We both sat straight up for ten minutes, and in
those ten minutes everything was deadly still. At

the end of them, we heard a cat's dolorous cry, away off, down below, somewhere.

"How can that cat have got in?"

"She isn't in; she's under the piazza, probably. She does go there. You'd better go to sleep, Lizzie." And Eylett laid himself down again, as men do when there isn't a fire nor anybody to shoot.

I knew I had better go to sleep; but I didn't fo. two good hours. By that time, I could hardly have declared that I had heard anything, it was so long ago, and I had so studied my impression to pieces, trying to match it to any possibility of causation.

Of course Eylett laughed at me in the morning; and of course I let him laugh, and didn't say anything till he got through. Women never do. Only when I thought he had had it out reasonably, I hushed him up as regarded the rest of the family. "Don't talk about it downstairs," I said.

He thought I wanted to be let alone on my own

account. It was not that. I wanted the fact let
alone. If it was not a noise, it was an experience.
That was what Mrs. Hand had called it. If you
have the experience, what difference does the noise,
or whatever else it may be, make, one way or the
other?

The next night I went to bed in a perfectly calm
and equable state of mind. I can positively affirm
that I expected nothing except to sleep. And I did
sleep, as I always do, instantly and soundly, after
my little read, which I always indulge in at night,
with a candle on my small book-table beside my
bed, in defiance of all old-time superstitions handed
down from the days of voluminous bed-curtains and
top-hamper, and absurdly repeated now, when we
lie down on our flat mattresses in their low French
boxes, with nothing combustible within a yard of
the light.

I slept my three or four early hours. I am glad

they are the hours of "beauty-sleep;" for they are
the only hours I am perfectly sure of. After that, I
begin to nap and dream, to wake, and think of
things, — the beans I meant to have told Adasha to
put to soak, the jam that must be scalded over, the
twist and buttons to be got for the tailoress who is
coming Thursday, then, being thoroughly roused,
to go round and regulate open windows, and cover
up the children.

It was, perhaps, about two o'clock when I was
again electrified into full and instant consciousness.
The same reverberating, radiating noise, ringing,
rattling, metallic, with a queer sound of struggle in
it, too, that suggested Pandemonium as one great
tin kettle, and all the little imps clawing frantically
to get out.

Then there came a bang. That woke Eylett.
Neither of us said a word, but both were instantly
out of bed and into dressing-gown and slippers.

We went into the great upper hall, and stood still. Everything else stood still, too. We could hear the old Willard clock ticking away composedly down in the dining-room, and not a breath or movement of anything else.

We went on, down between the rooms; as we went, there came winding up from somewhere, the eerie, weary, wandering wail of that uncanny cat.

Two doors moved their open cracks a little as we passed, and two noses were put forth.

"Marm! Sir!" cried Adasha Downe, in a tremulous whisper, "what was that racket?"

"What can have happened?" said Car Whapshare.

"Don't wake the children," whispered I. "We are going to see."

We went everywhere; up and down all the stairs, into the kitchen and pantries and out-rooms. We opened the side-door, and looked out into the starlight. Something black dashed out between Eylett's legs.

"I told you that cat was in," said I.

"Well, she's out," replied Eylett. "She couldn't have done it."

We found nothing to account for the clatter, not even a dipper or tin pan fallen down.

We went upstairs again, and encountered the noses waiting.

"What was it?" came the two whispers again.

"It doesn't seem to have been anything," answered Eylett.

"Marm!" said Adasha Downe, breathlessly, "that's awful!"

"No, it isn't," I retorted, with decision. "It's quite comfortable. Don't frighten the children."

In the morning I was dressed early, and went through the rooms upstairs with a vague feeling as if I might see by daylight where the sound had been.

There was a tin horse on the entry floor, lying peaceably upon its side, with that touchingly helpless and resigned expression that children's dolls and horses have in the cast-off positions in which little hands have left them; there was the usual litter of blocks and toys in the play-room, but nothing seemed as if it had borne part in any mystical orgie. The summer sun streamed in, and filled the chambers to the brim with cheer and splendor.

Coming out of the ell-room, I noticed the register-valve slipped slightly out of its place, and resting with one edge just over upon the floor. I pushed it back, and wondered who had moved it. I supposed Adasha must have lifted it out, in sweeping, to brush the dust from the spreading mouth of the pipe. I mentioned it to her when I went downstairs, and asked her to be careful. It would not do for the children to get an idea of its coming off. Adasha told me she had not "tetched" it. She didn't know it would come off. It was queer; but I supposed it "happened" somehow, and then I forgot all about it.

We had two still nights, and then in the third a rattle and a slam woke me up. I missed the reverberation, if it had occurred. In fact, I did not connect this with the other. It sounded like some one fumbling at a blind or lock, and then a sudden jar, as of blind or door flung back.

"It's burglars this time!" I whispered loudly in Eylett's ear. "I heard them trying something, and then it banged."

"Burglars don't bang," said Eylett, sleepily.

"There isn't any wind, and things don't bang themselves," said I. "You'd better get up."

So we had another promenade. It came to nothing, like the rest.

"Are we never to get any sleep in this house?" asked Eylett, in a melancholy way. "Don't hear anything more, Lizzie, if you can help it."

"No, I won't," I replied dutifully, keeping the rest of my thoughts to myself.

In the morning, before I went down the back-stairs to the kitchen to look after breakfast, stopping at the play-room, as I had a habit of doing, drawn by the pleasantness of the place where the children had been yesterday and were going to be to-day, and taking a glance at the sunshine and the toys that seemed conspiring a good time together, I saw that register off again,—really off, this time, an inch or two.

Could it have been that that banged in the night! I went back and called Eylett.

"Just look!" said I. "How do you suppose it came so?"

"Children," said he.

"No," I affirmed positively. "I found it so before; and I have watched. They never meddle with it; and, besides, it was not so at bedtime. We undressed them here. Do you suppose I shouldn't have noticed it?"

"Spirits, then," suggested Eylett, meekly, as driven to a logical end. "It's their style. Like their impudence."

"Pshaw!" said I, which was precisely what he wanted me to say.

For all that, the same night there was a greater din and rampage than ever; and the next morning there was the register fairly off and away, wheeled completely from the hole, and laid with nearly its entire circumference upon the carpet.

I called them all then,— Eylett, Caroline, and Adasha Downe. It was early. The children were only just waking up, and beginning to throw the pillows at each other, or to pull on stockings heel-side before.

"That ghost comes up the register-pipe," said Adasha Downe, solemnly, looking into the hole as into the mouth of the pit.

"And the ghost is"— cried I, with a sudden illumination.

"Never in this world!" broke in Eylett, catching my idea, and putting the extinguisher on before I had fairly shown its little blaze.

"Just lift that register," said he.

I put my hands under the two valves, an iron and a brass one. I suppose they weighed six or seven pounds. Could indeed a — well, the object of my suspicion — lift them up?

"I don't care," said I. "We'll see. I'll sit up this very night."

On the whole, however, when bedtime came, I decided to take that first nap, and trust to the usual reveille for warning. If I was right in my convictions, it would give me time enough. I am a light sleeper. I always hear the first stir.

I put a light in one of the ell-rooms, and set the door open upon the passage. I left another burning in my little sewing-room, back in its farther corner, and shaded so that it shone faintly out through the play-room.

A cross passage led over from opposite the head of the back staircase, between the rooms, to a linen-closet. Standing in this opening, or just down the first step of the staircase, one could command the whole scene of action, and nothing could pass in or out without observation.

I laid my dressing-gown and slippers in instant readiness. In fact, everybody else did the same; and we all slept, so to say, upon our arms; for everybody had petitioned, " Call me, if you hear anything."

Somehow, we were a little later that evening than usual; so that, with my ordinary and extraordinary preparations for the night, it was eleven o'clock, and the others were all asleep, when I was about to put out my own candle. Just as I had my hand upon the extinguisher, it began,— the noise.

That frantic, struggling, scratching, ringing, infernal sound, coming away up from depths below, and echoing everywhere.

"Quick! there it is already!" I cried to Eylett, and in the same moment was off myself. I darted in at the two doors on my way, and wakened the girls with one shake each.

"Don't be ten seconds, or else don't come!" I said, and hurried on. And in less than a minute we were all upon the spot, huddled, listening, lying in wait, in staircase and entry.

There was no doubt, standing there, where the sound came from. Up that long pipe from two floors below, it tore and grappled, grated and resounded ; came on, with pauses, higher and higher ; at last was on a level with ourselves. Then a fierce stirring and grinding, a seizing of hold and purchase. And then the valves clattered, as if pushed against, ineffectually, once or twice ; then, with a great hoist, they raised, swivelled, clashing round, and fell with an awful bang upon the floor.

That demoniac cat walked forth.

It was a positive fact. We saw it with our eyes. If anything in this story — my part or anybody's else — is embellished, it is not that.

"I told you so !" said I to Eylett.

And Eylett could not say a word.

We were all down cellar next morning, after our early breakfast, investigating ; and the more we investigated, the more we wondered.

Out of the brick dome of the furnace, high up, came the tin pipe that ran horizontally one-third or

more the length of the house, then up, twelve feet
perhaps, through the lower story and the two floors.

We opened the iron door of the air-chamber from
which the pipes radiated, and looked in. There
was only this one that started laterally; all the rest
sprung from the top. The furnace itself was built
close against a brick partition which divided the
cellar. A heavy padlocked door shut off the forward part, which had been Mr. Throop's wine-cel-
lar, and where all remained as he had left it.
Through some opening in the back, accessible only
from this locked division, must come the supply of
air to feed the furnace-chamber, and circulate in the
pipes. Through this, also, by ways known only to
herself, must have crept the cat, and likewise cir-
culated.

Into that dark, hollow space, up its rough-cast
sides, — into the small, utterly obscure aperture,
along those twenty feet of mystery and uncertainty,

—one would think this was exploit and marvel enough; but up that twelve feet perpendicular, with nothing but the lapping of the tin sheets to claw by, and the bracing of her body between the narrow sides! Beyond that, the closed register at the top! What sort of faith, or instinct, or impishness, was it that led her on? We stood in utter, awed bewilderment. It was almost stranger than a ghost.

One thing was certain, we could not let the play have a run of a hundred nights. Something must be stopped up, or come down.

"The hole in the furnace," suggested Caroline.

"We can't get at it."

"Nail something over the register."

"Then we should have the noise, all the same, and the poor cat would have to tumble twelve feet, and crawl twenty backward. She deserves better for her smartness."

"Unhitch the pipe."

"We can't have workmen into the house, or alter anything."

"I'll do it myself," said Adasha Downe. And she straightway ran up the cellar staircase, beside which passed the pipe, and laid brave hold.

A neck of iron was set in the brick-work of the furnace, around which fitted the tin sheet. Adasha pulled and pulled; but what could she do with twenty or thirty feet of metal cylinder, and years of rust? Eylett stood still considering, while she strove unheeded. Then he went and got a hammer and a chisel. Then I climbed up on a barrel, on the other side of the pipe to where Adasha was. Caroline took the children up the staircase, and kept them there peering down at us in a little eager heap from its head.

Eylett hammered and loosened, and we pulled. We all pulled. Eylett twisted; and presently, all of a sudden, some weak joint gave way above, and, at the same moment, the neck yielded, and — crash! down came the whole thing, revenging itself upon us by its compliance.

"O mamma! mamma!" cried out Robbie; for I and my barrel had tumbled down. Adasha seated herself very hard upon the stairs.

"Are you hurt, Lizzie?" cried Eylett, coming in a hurry.

No. Nobody was hurt. Only the pipe was separated in two or three places, the air was full of dust, and we felt as if we had pulled half the house down.

"Phew! phew!" said Eylett; and brushed his hands against each other, and looked at the wreck.

He lifted a long piece, and set it up on end against the wall. Out of it, as he did so, fell a great deal more dust, and other things which we perceived as the dust subsided. A great many pins, — of course; an old piece of black comb; a red chessman; nutshells; a brass thimble; hair-pins; corks; a handful of coppers that probably used to roll out of Zerub Throop's trousers-pockets when he pulled them off; in the midst of the heap, something round and bright, like a silver ball.

The children — little wreckers that they always are — were down again by this time, notwithstanding remonstrances. They couldn't help it; they kept minding, and going up, and irresistibly gravitating down again, in little sprinkles, one and two at a time.

Robbie pounced upon the shining thing.

"Oh, I speak for that! Is it a silver dollar, mamma?"

Poor Robbie had heard traditions of silver dollars, earned and saved up in his father's childhood; but his little experimental knowledge stretched not beyond the days of scrip.

"Oh, no!" I said, foolishly. "That isn't a dollar. It isn't anything."

"Not anything, mamma? Why — why — here it is!"

"I'll tell you what it is," said Blossom, standing daintily on the stairs out of the dust, with her fresh piqué frock, and her little white stockings. "It's a fairy ball, and Miss Whapshare will tell us a story about it."

"So I will," said Car, seizing her opportunity. And she got them all away, up out of the cellar.

What she told them I don't know, — about fairy balls that opened, and had wonders inside; and fairy balls that only rolled and rolled and rolled, and led people along through forests and among mountains, and out into some paradise perhaps, of elf-land, at last; but when I had changed my dusty dress, and washed my face and hands, and seen Eylett brushed up and off to the train, I found them

all together in the play-room; Car, with the ball in
her hand, and Robbie and Blossom beseeching her
to open it.

"Then it will be spoiled," she said, "if it isn't an
opening ball. I think it is a rolling one. It must
have rolled down the register. Who knows where
it will roll next?"

Behind me up the stairs, in a fashion of privilege
she had taken, came suddenly Sarah Hand.

And, of course, then came the story, — all about
the cat, and the pipe, and the ball.

"You see a great tin piece of the house came
down when they pulled," said Robbie, "and broke;
and everything came out, — cents and pencils and
everything."

"Droppin's and sweepin's," said Sarah Hand.
"That's how they came there."

"Not my fairy ball," said Robbie. "That rolled
itself. Nobody knows where it rolled from. Way

down and down, and over and over, and all through the world."

"I'll tell you where it rolled from," said Sarah Hand, taking it up. "I remember it. It's one of the things that used to lay round on Zerub Throop's table. I know 'em all by heart; the things I used to turn over and dust, and put back careful. I noticed that, because it looked as if there might be something did up in it. He fixed it his own self one day after dinner. I recollect the day too. 'Cause Miss ——, he'd had a visitor, and we'd had a talk. I s'pose he was jest settin' thinkin'. It's kinder awful, comin' across things so, after folks is dead and gone."

And Mrs. Hand laid back the ball on Caroline Whapshare's lap.

Caroline took it up as if by a sudden impulse, and picked out one edge of the folded foil. A little tremor passed over her.

"What is the matter?" said I.

"Nothing. I shivered, I don't know why."

"Um!" said Mrs. Hand, and looked solemn.

"I think that might as well be unrolled, and done with, now the story is told," I said briskly; for the children's eyes were getting big. "We shall be having little nightmares of the ball travelling about, if we don't take care."

Then Caroline turned back corner after corner, edge after edge, until two ends were opened out. It was no longer a ball, but a little roll. There was something in it.

Paper, — written paper, folded and coiled.

"I feel as if it were a secret," said Caroline, as the last doubling of tin-foil fell away, and left it in her hand.

"Perhaps it is. But there is nothing hidden" —

I stopped. Car had got the paper open, had

given one glance at it, and every bit of color had flashed instantly out of her face.

" Mrs. Bright! What does it mean? "

And poor little Caroline burst out crying. That saved her from fainting away.

I took the creased and curled-up scrap.

" *For value received of Miles Whapshare, in the year one thousand eight hundred and forty-five, I promise and direct to be paid to Mrs. Miles Whapshare, widow of said Miles Whapshare, or her heirs-at-law, six months after my decease, or on the présentation of this paper to my executors at any time within five years from such decease, Thirty-five thousand dollars.*

" ZERUBBABEL THROOP."

I turned it over.

" *October 19th, 1866.*

" *Left to Providence.*

" *Payable to order; that is, on turning up.*"

We sent for Rufus Abell and for Dr. Plaice.

It was all quite plain, and strong; as strong as it was queer.

"This is the thing that was provided for," said Rufus Abell, just as unmoved as if he could possibly have expected it. I suppose Mr. Abell had got over surprises long ago.

Arthur and Caroline went home together to tell Mrs. Whapshare.

I watched them go down the hill in the sunshine, gathering it, as it were, around and after them, to carry down in one great golden rush into the corner house that had been full of little crowding clouds of care so long. I thought of that bit of creased-up paper in Rufus Abell's wallet, and how it would go to probate with the will, and settle everything, and how strange, and changed, and wonderful it all was. And I bit my tongue to try if I was awake; and then I turned round and said to Mrs. Hand : —

"To think it should all be by means of that cat!"

"It's very well," said Mrs. Hand, with slow significance, "to lay it all off on to her. But what possessed the *cat?* It's like the pigs in the New Testament. If — a ghost — wanted something — out of a register-pipe, — he might very likely need some sort of a cat's-paw to help hisself with."

Was it a cat, or was it a ghost, or was it simply Providence? It was the question left on our minds. We thought, humbly and honestly, that it might be all three. We put this and that together that we had learned, and we believed it just possible, among the mysteries, that Zerub Throop had at last "come across Providence," and had been set to work perhaps with such links and agencies on earth as he had established for himself.

At any rate, the Ghost Story and the Cat Story

got so mixed up and merged, that they were never popularly disentangled.

We could never get any other girl than Adasha Downe to live with us at Throop Hill, though we came there three summers.

"The owld man might ha' left somethin' else that needed seein' after; who knows?" the Irish said.

Caroline Whapshare and Arthur Plaice were married in September. Mrs. Whapshare gave them five thousand dollars.

"There would be that," she said, "for each of the children, and the same for her own part. They should have their share as they came to want it. She'd done waiting enough herself for the whole family."

Miss Suprema Sharpe had a kind of congestive fever that fall. She took cold at the wedding. But the doctor did not think that was quite the whole of it. There was a feverish fulness that must determine somewhere, — a greater pressure than the ordinary circulation could carry off.

A ghost-story, a wedding, and a fortune, — what they did with it, and how they behaved about it, — all this, you see to come right in here, like an avalanche, at the corner, to be thoroughly sifted and discussed, and realized and criticised; well, it could not have gone much harder with Suprema Sharpe; and if you knew her as we do, Dutton, you would understand.

It isn't a matter to make fun of, though, and I wouldn't have you think I do. It's an awful fact, and a solemn retribution. There is such a thing as a vacuum in heart, or brain, or life, by which the surrounding atmosphere has to press in with fifteen uncompensated pounds to the inch. And that is the way the burden of everybody's else affairs comes down at last upon the Sharpes.

10

That couldn't have been in Dante; could it, Dutton dear?

But if Dante had come after Kepler and Newton — and a few other folks — I guess it would have been.